William R. Johnson, CRA, FAHRA, MBA

Service Delivery vs. Service Excellence

Opposing Forces during the Patient Encounter and How to Overcome Them

ISBN: 978-1-304-49770-3 (sc)
ISBN: 978-1-4834-0686-2 (e)

Library of Congress Control Number: 2014900034

Lulu Publishing Services rev. date: 1/10/2014

To my parents, Christine "Teeny" Johnson and George "Eggie" Johnson, thank you for adopting me and instilling within me values that have served me well throughout my life. You were the best parents and role models a son could ever hope for. A big part of you both is found within the pages of this book. I love and miss the both of you.

To my wife, Vicki, your patience, love, and support have proven to be the lifeboat of our marriage and relationship. Thank you for always being there for me and encouraging me to pursue this book. I hope we have many more years of love, life, and laughter together. You are an incredible woman, and I love you.

To Paul, I am so proud of you and admire the man you have become. You have proven to me the truth in that old Albert Einstein poster that hung on your bedroom wall years ago that read, "Brilliant spirits have always encountered violent opposition from mediocre minds." Continue to light up the world and pursue your passions in life. I love you, Daremaster Spoodle.

CONTENTS

FOREWORD

Many of us in the health-care industry work extremely hard on getting patient satisfaction right only to have patients tell us with regularity, "You got it all wrong!" There are times our patient satisfaction results take an up-and-down roller-coaster ride and in turn we drive ourselves crazy trying to identify the cause or causes of the fluctuations.

Trying to fully understand the *why* of wildly fluctuating patient satisfaction results and its remedies can be compared to trying to identify, capture, and eradicate a poltergeist. The apparition appears out of thin air and then, without explanation, fades away. A few months later a different type of ghost returns to torment us once again. It's maddening! If you are a follower of Erik Wahl (www.theartofvision.com), you'll understand when I say, "It's enough to make you sniff a Crayola crayon all day."

In a zealous quest to achieve best-of-class service excellence organizations risk becoming hostage to a happiness number. It's a hard lesson to learn that satisfaction results cannot be managed. Only the patient experience can be managed. The great news is if the patient's experience is managed well the numbers come naturally and validate we are finally getting it right.

So why does it feel like we are set up for service excellence failure even before the first patient of the day walks through the door? It's because in many ways we *are* set up for failure. The very nature of how we deliver clinical, diagnostic, and therapeutic service inadvertently creates obstacles in our ability to achieve world-class service excellence. We will talk more about that paradox later in the book.

Let's begin by making sure we are all on the same page as to why patient satisfaction matters. At the writing of this book the health-care industry is in turbulent times, especially when it comes to reimbursement. The payment model has quickly changed from "volume equals revenue" to "outcomes equal revenue." The Centers for Medicare and Medicaid Services (CMS) value-based purchasing (VBP) model places revenue at risk based upon performance in both clinical measures and the patient experience. Providers who perform poorly in either or both categories lose reimbursement. The primary reason for wanting to deliver an incredible patient experience, however, should not be just about money.

Many health-care systems and providers focus a lot of energy into capturing and maintaining market share. Being able to increase or maintain market share is often strongly correlated to highly satisfied patients who, over time, become extremely loyal.[1] High market share in many ways equates to a more predictable and steady revenue stream.[2] Loyal patients become a sales force and help grow business through word-of-mouth referrals and personal testimonies, the strongest sales tools in existence. Aspiring to achieve superior service excellence should be driven by more than simply trying to dominate a given health-care market.

The true value of patient satisfaction is what it does for the patient. Providing the patient with an incredible experience not only helps shape the patient's perception of higher quality care but also creates a trusting and caring relationship between the patient and the care provider. That trusting and caring relationship positively influences the patient's willingness to become engaged and take ownership in their own health outcomes. It influences the patient's willing compliance, for example, with a provider's home-care instructions.[3] Patient satisfaction, which is primarily a subjective measure of emotion and sentiment, is another tool that providers have in their arsenal to influence clinical outcomes.

For example, a study conducted by William Boulding, PhD, and associates and published in the *American Journal of Managed Care* demonstrated higher hospital-level patient experience scores (i.e., Hospital Consumer Assessment of Health-Care Providers and Systems domains *Overall Rating*

---

1   James L. Heskett, Thomas O. Jones, Gary W. Loveman, W. Earl Sasser, Jr., and Leonard A. Schlesinger, "Putting the Service-Profit Chain to Work," *Harvard Business Review* 72, no. 2 (March–April 1994): 164—174.

2   Andrew N. Garman, Joanne Garcia, and Marcia Hargreaves, "Patient Satisfaction as a Predictor of Return-to-Provider Behaviors: Analysis and Assessment of Financial Implications," *Quality Management in Health Care* 13, no. 1 (Jan.–March 2004): 75–80.

3   Edward E. Bartlett, Marsha Grayson, Randol Barker, David M. Levine, Archie Golden, and Sam Liffer, "The Effects of Physician Communications Skills on Patient Satisfaction; Recall, and Adherence," *Journal of Chronic Diseases* 37, no. 9–10 (1984): 755–64.

of Hospital and *Discharge Planning*) were independently associated with lower thirty-day readmission rates for acute myocardial infarction, heart failure, and pneumonia admissions.[4] It turns out patient satisfaction is more than hugs, kisses, and warm fluffy stuff. It is another way providers effectively work toward patient-centered wellness care and positively influence the health outcomes of their patients.

So let's jump in and examine the dynamics between service delivery and service excellence and gain a better understanding and appreciation for the negative influence service delivery models sometimes have on the patient's experience. By better understanding how the logistics of service delivery set the stage for service excellence roadblocks we can develop strategies to overcome the obstacles we ourselves create and ultimately deliver incredible patient experiences.

---

4   William Boulding, Seth W. Glickman, Matthew. P. Manary, Kevin A. Schulman, and Richard Staelin, "Relationship between Patient Satisfaction with Inpatient Care and Hospital Readmission within 30 Days," *American Journal of Managed Care* 17, no. 1 (2011): 41–48.

CHAPTER 1

Left-Brain vs. Right-Brain Orientation

To better understand how service delivery and service excellence oppose each other during the patient encounter, let's begin by reviewing the theoretical functions of the left and right hemispheres of the human brain.

Research suggests the brain's neocortex, which is divided into left and right hemispheres, not only gives us the ability to process abstract thoughts, words, symbols, logic, and time but also serves as the home of emotions, subjectivity, and creativity.[5]

It is theorized that each hemisphere of the brain, referred to as the left brain and the right brain, serves specific functions. The left brain is the source of analytical thought, organizational skills, complex problem solving, and objectivity, while the right brain is the source of emotions, creativity, intuition, and subjectivity.

The very essence of left- and right-brain functions actually becomes a source of disharmony between the patient and the provider in a typical health-care setting. Let's explore why that happens.

In the typical health-care setting, the operational logistics of service delivery is created based on left-brain-oriented protocols, techniques, processes, policies, and procedures. Service delivery is calculated, analytical, and purposeful. Delivery processes are intentionally created this way to ensure predictable outcomes in areas such as quality and safety.

When one is deciding how service will be delivered, the patient is

---

5   George Kohlrieser, *Hostage at the Table: How Leaders Can Overcome Conflict, Influence Others, and Raise Performance* (San Francisco: Jossey-Bass, 2006), 5–6.

usually not the center of focus. Rather, service delivery models established by the health-care system, hospital, or department are often created to ensure consistent compliance with federal and state regulations, the demands of those paying for the services (private and government payers), and the provider's own interests (efficiency and cost savings).

Let's think about what it takes to deliver a diagnostic test, such as a computerized axial tomography (CAT) scan of the chest, to a patient. Long before the patient receives the service, the provider of the service (medical imaging center or medical imaging department) must obtain some type of an order or written directive from an individual legally authorized to direct the test or treatment be performed. We often call this legally authorized individual a doctor or clinician. The information contained within the order must justify the need for the CAT scan (known as medical necessity). In most cases, the payer, who will reimburse the service provider, requires preauthorization long before the service is delivered.

Once the order is obtained and reimbursement has been arranged, other hurdles need to be cleared, such as privacy and confidentiality forms being signed and copies of insurance and identification cards collected. Oh yes, let us not forget about the most important step of all, the Holy Grail of all left-brain-oriented service delivery models, the collection of the copayment.

The left-brain-oriented service delivery process goes on and on, until the patient finally receives the CAT scan. Unfortunately, the health-care professionals who are a part of this service delivery process and have the best of intentions often fail to meet the most basic need of the patient, namely, treating the patient like a human being and not another widget on the assembly line.

Every day in the United States, millions of patients move through their health-care encounters and receive care from the most dedicated medical professionals working in the most incredible health-care organizations on the planet. So how could patients possibly become dissatisfied with what they receive during their health-care journey when they are surrounded by highly skilled clinical experts using the most advanced medical technology available in the world? The answer lies not in what is being delivered or who is delivering it but *how* it is being delivered.

Patients become dissatisfied with their health-care experience because most can only receive and evaluate clinical services from a right-brain, emotional perspective. They do not have the ability to assess clinical delivery and competency from a left-brain perspective. These patients experience their entire health-care journey through the right brain. Since health care is delivered from a left-brain perspective and the patient receives it from a right-brain perspective, there are opposing forces at work from the first moment the patient experience begins.

Don't misunderstand me. Technical and diagnostic skills are important, but patients do not assess those as quickly as they do the quality of the interactions among the physicians, staff, and themselves.[6] You see, patients and families hold in their minds a mental picture of how a person should be treated, and that becomes the standard by which they judge their health-care experiences.[7]

In 2012 Lonnie Hirsch, cofounder of Healthcare Success Strategies, reported that more than 60 percent of patients define the patient experience in a hospital by how well they are treated as a person, more so than the medical treatment itself.[8]

When patients talk about quality of care, they frequently refer to the *quantity of caring* they received during their health-care journey. Bottom line is patient satisfaction is a measure of emotion, or sentiment, about their health-care experience. These feelings are by-products of their encounters with staff and physicians. Patient satisfaction many times comes down to the patient's feelings about the behaviors of the providers within the service delivery process.

American author and poet Maya Angelou sums it up: "I've learned that people will forget what you said, people will forget what you did, but people will never forget how you made them feel."[9] Nothing is truer when it comes to the patient's health-care encounter.

In the left-brain-dominated world of health-care, providers often lose sight of what matters most to patients, and that is exhibiting affective qualities such as respect, compassion, patience, friendliness, and advocacy.

6   Susan Keane Baker, *Managing Patient Expectations: The Art of Finding and Keeping Loyal Patients* (San Francisco: John Wiley & Sons, 1998), xi.

7   Fred Lee, *If Disney Ran Your Hospital 9½ Things You Would Do Differently* (Bozeman, MT: Second River Healthcare Press, 2004), 10.

8   Lonnie Hirsch, "15 Best Practice Reasons Professionals Care about Patient Satisfaction," *PatientExperience.com,* November 20, 2012, http://patientexperience. com/15-practice-reasons-professionals-care-patient-satisfaction/.

9   Bob Kelly, *Worth Repeating: More than 5,000 Classic and Contemporary Quotes* (Grand Rapids, MI: Kregel Publications, 2003), 310.

Patients perceive those as caring behaviors.[10] When all providers display caring behaviors, it creates a caring culture, which envelops the patient during his or her health-care journey.

In the next chapter, we look at the importance of creating a patient-perceived caring culture and how a caring culture becomes a primary driver of the patient's overall satisfaction.

---

10   Wendy Leebov, *Essentials for Great Patient Experiences: No-Nonsense Solutions with Gratifying Results* (Chicago: Health Forum, 2008), 15–20.

CHAPTER 2

The Importance of a Caring Culture

In the health-care setting, a left-brain-oriented service delivery process is critically important for the welfare of the patient and cannot be abandoned. The processes, policies, and procedures that make service delivery possible also ensure the patient's safety and help create quality outcomes. Left-brain-oriented service is not the enemy during the patient encounter.

Patient dissatisfaction comes from the imbalance between the overbearing and dominating left-brain service delivery processes and the absence of right-brain caring behaviors exhibited by staff and physicians. If service excellence is to be achieved, everyone who comes in contact with the patient, no matter how brief the encounter, must learn to consistently exhibit behaviors that create an emotional connection with the patient and demonstrate caring.

In a typical health-care setting, the average patient is unable to fully understand and appreciate the complexity or functionality of the technology used in the delivery of service. The patient has very little knowledge of the staff's clinical or technological competency. The patient, however, fully understands and can quickly evaluate the staff's interpersonal skills during the health-care experience. Being treated like a human being and not a diagnosis matters most to the patient.

In the example of patient John Doe's CAT scan, the availability of advanced technology and the competency of the staff using that technology are not just his expectations; they are nonnegotiables Mr. Doe brings to the encounter. He fully expects the most advanced technology operated by the most skilled and competent staff to be used in the delivery of his care—without exception. What truly matters to Mr. Doe is to be treated with empathy, caring, compassion, and respect during his health-care journey.

Dr. Robert D. Sheeler of the Mayo Clinic understands the importance of a caring culture in health care. When he works with medical residents, he stresses the importance of demonstrating caring to students and patients alike. "Being able to communicate and motivate both patients and learners," he states, "is summed up in the phrase I constantly use when teaching medical students: they [patients] have to know that you care before they care about what you know."[11]

Health-care professions often overlook the power of a caring culture or a caring gesture. Simply touching the patient's hand and saying, "I'm here" can be one of the most important moments in a patient's health-care journey. Just ask Marcus Engel about the power of those gestures.[12]

The power of a caring culture and the influence it has on clinical outcomes has been documented in research for many years. In 1984 researchers Edward E. Bartlett and associates investigated how caregiver communication skills impacted patient satisfaction, as well as patient adherence to his or her medication regimen. In this study, researchers found medication adherence was influenced by patient satisfaction, which was determined by the quality of the physician's interpersonal skills. In fact, interpersonal skills were more important in determining patient outcomes than patient education.[13]

Edwin Boudreaux and Erin O'Hea analyzed fifty of the best-designed and -implemented research articles for patient satisfaction in emergency departments (ED). They found the predictor domain most strongly associated with ED patient satisfaction was interpersonal interactions with ED providers. Interpersonal interaction included interpersonal mannerisms and perceived humanitarian behaviors, or simply put, bedside manner.[14]

---

11   David Matsumoto, Mark G. Frank, and Hyi Sung Hwang, *Nonverbal Communication: Science and Applications* (Los Angeles: Sage Publications, Inc., 2013), 240.

12   Marcus Engel, The *Other End of the Stethoscope: 33 Insights for Excellent Patient Care* (Orlando, FL: Ella Press, 2006), 23.

13   Edward E. Bartlett, Marsha Grayson, Randol Barker, David M. Levine, Archie Golden, and Sam Liffer, "The Effects of Physician Communications Skills on Patient Satisfaction; Recall, and Adherence," *Journal of Chronic Diseases* 37, no. 9–10 (1984): 755–64.

14   Edwin D. Boudreaux and Erin. L. O'Hea, "Patient Satisfaction in the Emergency Department: A Review of the Literature and Implications for Practice," *Journal of Emergency Medicine* 26, no. 1 (2004): 13–26.

In 2001 Kris Kipp reviewed how implementing nursing caring standards in an ED increased patient satisfaction ratings. In her study, Kipp points out that caring is a core characteristic of the nursing profession, not just clinical competency. Kipp found that ED patient satisfaction in regard to "care and concern by nurses" increased 6.6 percent after nursing caring standards were implemented as standard practice in the ED.[15]

In another study, researchers investigated factors that influenced the frequency of return visits by the homeless to the ED who regularly visited the ED for vague and nonmedical reasons.[16] The researchers randomized patients into two groups—compassionate care plus standard medical care or just standard medical care. Compassionate care consisted of a person who sat with the patient, chatted with him or her, or got the patient a cup of water, blankets, or other nonmedical comfort measures.

It was found that the homeless patients in the compassionate care group returned less frequently to the ED. It was also found that once the patient's need for interpersonal contact, or socialization, had been more fully met they returned to the ED less frequently. Another example of how a caring culture influences health-care outcomes and utilization.

The evidence is endless. This book is not large enough to review all the available research that validates the power of caring behaviors and a caring culture in a health-care setting. The bottom line is patient satisfaction, and the patient's perception of the quality of care improves when the patient experiences caring behaviors from their providers. A caring culture matters, and it is one of the things that matters most to the patient!

In the next two chapters we will explore the behaviors, both nonverbal and verbal, all members of the service delivery team must habitually exhibit to create what the patient perceives as a caring culture.

---

15   Kris M. Kipp, "Implementing Nursing Caring Standards in the Emergency Department," *Journal of Nursing Administration* 31, no. 2 (2001): 85–90.

16   Donald A. Redelmeier, Jean-Pierre Molin, and Robert J. Tibshirani, "A Randomized Trial of Compassionate Care for the Homeless in an Emergency Department," *The Lancet* 345 (1995): 1131–34.

CHAPTER 3

Behaviors, both verbal and nonverbal, can reflect not only the beliefs and values of an individual but also the beliefs and values of a group. Overt behavior of a member of a group is determined by both the cultural predisposition, which are the perceptions, thoughts, and feelings that are patterned by the group, and the situational contingencies that arise in response to the external environment.[17] Simply stated, the actions of an individual is greatly determined by the norms of the group (culture) they are a member of and the demands of the situation at hand.

Cultural norms are shaped, established, and reestablished over time. The person who ultimately takes ownership and responsibility for the culture of a group, along with taking responsibility for the performance and behaviors of the group, is referred to as the 'leader'. In the workplace the leader is often formally recognized through some sort of a positional title such as supervisor, manager, director, vice president, or CEO.

If a different level of performance is needed from a group it is necessary for the values and beliefs (norms) of the group to be modified before a change of the group's performance will occur. The responsibility of changing the values and beliefs of a group falls to the leader. Changing a group's culture should be a leader driven, top-down phenomena and not bottom upward.

Cultural change is extremely complex and this entire book could be

---

17   Edgar H. Schein, *Organizational Culture and Leadership*, 3rd ed. (San Francisco: Jossey-Bass, 2004) 19.

dedicated to that single subject. For simplicity sake let's take the position that the leader has a strong influence on the nature of the group's culture and the leader is greatly responsible for shaping the norms of the culture. If the leader does not take the initiative to shape the culture and to establish the norms, values, and behavioral expectations of the group then the group will create the culture themselves. Generally, if the leader does not manage the culture, the culture will eventually manage the leader.

The leader must role model the values and beliefs he or she wants the staff to emulate. The leader must hold staff accountable for their behaviors which should reflect the desired cultural norms. Converting an uncaring culture or semi-caring culture to one that is caring will only be achieved through strong leadership. The leader must have the passion and determination that the group will deliver what matters most to the patient and that is staff exhibiting caring behaviors within a caring culture.

Nonverbal behaviors are powerful in communicating the values of the individual and the culture. Nonverbal behaviors not only give insight into an individual's temperament and intentions but also have specific meaning in a given context or situation in which they are being used.[18] That being said, let's identify the basic nonverbal behaviors everyone within the service delivery group must consistently exhibit to create a culture the patient perceives as caring.

Keep in mind that everyone within the care delivery team is on stage and under scrutiny every minute of every day. Caring behaviors must be habitually displayed by each member of the group while in front of patients, the patient's families, other visitors, and especially coworkers. Caring behaviors are not just shared or displayed with only the patient. Caring behaviors must be displayed at all times in front of everyone if caring is to become the cultural status quo.

Nonverbal behaviors are important and often carry more weight than verbal behaviors. The phrase, "Actions speak louder than words," is so very true. In general, 55 percent of the way most of us communicate is nonverbally.[19] The patient's first impressions of the culture and the people working

---

18   Lisa Slattery Rashotte, "What Does That Smile Mean? The Meaning of Nonverbal Behaviors in Social Interaction," *Social Psychology Quarterly* 65, no. 1, (March 2002): 92–102.

19   Jeff Thompson, "Is Nonverbal Communication a Numbers Game," accessed August 20, 2013, http://www.psychologytoday.com/blog/beyond-words/201109/is-nonverbal-communication-numbers-game

within are shaped in a large part by the nonverbal behaviors exhibited by the staff.

Impressions become a filter through which the patients interpret, evaluate, and measure their experiences during their health-care journey. Impressions matter regardless of whether it's the first impression or the last. Ultimately all the impressions come together in the patient's mind to create a *cultural* impression. Here is a simple example of how impressions are formed and the power they carry:

a.   The patient takes in information, noticing the body language, behavioral responses, and spoken language of the staff encountered and observed.

b.   Based upon this information, the patient forms impressions and assumptions about the culture and people within and makes decisions about what the staff are like and how they will behave in the future.

c.   The patient then sees everything through this impression filter. Unfortunately the patient looks for information that is consistent with his or her cultural impressions and ignores behaviors that do not fit his or her cultural impression.[20]

That is why if the patient's first impressions of a department or hospital go badly it is extremely difficult to change the cultural impression formed in the patient's mind.

There is simple nonverbal behavior that staff, including physicians and senior leadership from the CEO down, must consistently exhibit to reflect a caring culture. Everyone is a part of the culture and the patient expects everyone within the culture to behave in a caring manner.

Notably, there are sometimes cultural differences regarding the use of both nonverbal and verbal behaviors. Generally nonverbal behaviors like eye contact and smiling are accepted as standard interpersonal skills. There are, however, certain cultures that have different expectations regarding certain nonverbal behaviors. There are several excellent resources that can assist health-care providers in better understand the values and expectations of differing cultures. One such easy-to-use resource comes from the

---

20   Ann Demarais, Valerie White, *First Impressions: What You Don't Know about How Others See You* (New York, NY: Bantam Books, 2004) 17.

Joint Commission Resources and can be purchased from their website www.jcrinc.com. When on the website search 'cultural sensitivity'.[21]

Let's begin with reviewing the importance of *eye contact* during the patient encounter. Making eye contact is an important step not only during the onset of the patient encounter but also during the encounter's duration. Socially we scan the faces of each person we see and attempt to make brief eye contact in the hopes of better understanding the state of mind or demeanor of the individual. It is human nature for us to behave in this way. When face-to-face interaction unfolds between people moderate eye contact is used to build initial nonverbal rapport. Rapport occurs when two people attempt to get 'on the same page,' and it generally greases the wheel of social interaction.[22] During a normal interaction researchers discovered we make eye contact 45 to 60 percent of the time.[23]

Eye contact is the way we confirm each other's existence. It is also the way we demonstrate interest in one another. If one person purposely avoids eye contact with another it can be interpreted as disinterest, disengagement, and even antisocial behavior. Having eye contact with the patient is an important part of building a perception of interest, empathy, and caring.

Research suggests that eye contact between the health-care provider and patient helps build the perception of empathy.[24] Nonverbal cues like eye contact help providers communicate respect for the patient along with reinforcing the provider's credibility and enthusiasm.[25]

Nonverbally eye contact alone is not enough to build the perception of caring. It cannot be the only nonverbal we bring to the table. We must add another nonverbal cue in an attempt to build the perception of pleasantness and a positive atmosphere. We must *smile* when we are making eye contact.

---

21   Geri-Ann Galanti, *Cultural Sensitivity: A Pocket Guide for Health Care Professionals*, 2nd ed. (Oakbrook Terrace, Illinois: Joint Commission Resources, 2012) http://store.jcrinc.com/cultural-sensitivity-a-pocket-guide-for-health-care-professionals-second-edition.

22   Matsumoto, *Nonverbal Communication: Science and Applications*, 90.

23   Demarais, *First Impressions: What You Don't Know about How Others See You*, 62.

24   Enid Montague, Ping-yu Chen, Jie Xu, Betty Chewning, and Bruce Barrett, "Nonverbal interpersonal interactions in clinical encounters and patient perceptions of empathy," *Journal of Participatory Medicine* 5 (August 14, 2013).

25   Jennifer S. Morse, "Improving Patients' Satisfaction through Positive Communication," *Cataract & Refractive Surgery Today* (April 2009): 102.

By using both eye contact and smiling we are nonverbally communicating to another person, "I see you, you see me, and it's all good."

Smiling is like a magnet. It pulls people into the interaction. Patients become more relaxed and participate more willingly in the interaction if the care provider is nonverbally communicating pleasantness through a smile. Like eye contact, smiling is another way rapport is created between individuals.[26]

Patients appreciate a health-care provider who presents him or herself in an appropriate manner, and that includes the provider having a smile on their face.[27] Smiling is also a great way to help break the ice with the patient and initiate personal interaction.[28]

Socially smiling is a product of both nature and nurture.[29] As infants we inherently (nature) smile to express pleasure. As adults society teaches us (nurture) to smile in certain social situations. For example, we are taught to smile during conversations in an effort to nonverbally communicate politeness during the interaction.[30]

As the interaction between provider and patient continues to unfold body position takes on importance. The provider must nonverbally express interest in having an interaction with the patient. Engagement and having a sincere interest in interacting with another person must be clearly communicated nonverbally before the first word is spoken. In order to achieve this goal, the provider *must face* the patient or what is referred to as *pointing your heart.*

The heart is one of the most vulnerable parts of the body. When the

---

26   Ken J. Rotenberg, Nancy Eisenberg, Christine Cumming, Ashley Smith, Mike Sing, and Elizabeth Terlicher, "The contribution of adults' nonverbal cues and children's shyness to the development of rapport between adults and preschool children," *International Journal of Behavioral Development* 27 no. 1 (2003): 21–30.

27   Marianne M. Lill, and Tim J. Wilkinson, "Judging a book by its cover: descriptive survey of patients' preferences for doctors' appearance and mode of address," *British Medical Journal* 331 (2005): 1524—1527.

28   Wendy Leebov, and Gail Scott, *Service Quality Improvement: The Customer Satisfaction Strategy for Health Care*, (Lincoln, NE: Authors Choice Press, 2007), 44—45.

29   Susan Jones, "Nature and Nurture in the Development of Social Smiling," *Philosophical Psychology* 21, no. 3 (June 2008): 349–357.

30   Zara Ambadar, Jeffrey F. Cohn, and Lawrence I. Reed, "All Smiles are Not Created Equal: Morphology and Timing of Smiles Perceived as Amused, Polite and Embarrassed/ Nervous," *Journal of Nonverbal Behavior* 33, no. 1 (March 1, 2009): 17–34.

provider is ready to initiate the patient encounter the provider must be willing to become somewhat vulnerable. Making eye contact, smiling, and then pointing the heart demonstrates trust and vulnerability during a social encounter and requires confidence and courage. By opening up our vulnerability we extend a great deal of trust to the other individual in the hopes they will open up and become vulnerable in a similar manner. Someone, however, must go first and that must be the provider because of the enormous weight he or she carries during the patient encounter.

Think about the last time you had a meaningful and perhaps emotional conversation with another individual. More than likely you would describe that conversation as having a heart-to-heart with the other person. Now imagine how a patient might feel when he or she is interacting with a provider who does not make meaningful eye contact, will not smile, and does not face them during a conversation. Obviously the patient will not have positive feelings about that provider.

Eye contact, smiling, and facing someone during an encounter is the way in which positive, healthy, caring, and productive interactions take place. In order to understand how the power of these three nonverbal behaviors have in your personal life please take a moment to participate in the exercise that follows.

Exercise: Think of two people in your personal life. Person number one is someone you enjoy being with and look forward to spending time with. When you are with person number one you have great conversations and have positive feelings about the interaction and relationship. When the encounter is over you look forward to seeing that person again in the future. Can you picture in your mind the face of person number one? Good. Now let's think about person number two.

Person number two is the person in your life that just strains your very last nerve. You do not look forward to spending time with this person and in fact you try to come up with excuses to avoid being with him or her. When you have to be with person number two you may notice your body is tense, your toes may be curled, or perhaps you find yourself gritting your teeth. The tension between yourself and person number two could be cut with a knife. When it is over you feel fatigued, maybe even exhausted and leave with negative feelings about the encounter. Can you picture in your mind the face of person number two? Good.

Of the two people you pictured in your mind in the exercise above which one, during your time with them, do you have frequent eye contact with, find yourself frequently smiling, and purposely facing them during conversations and which person do you find yourself avoiding eye contact with, frowning, and turning away from during conversations?

This is how we behave as human beings and this is how we nonverbally communicate our feelings about each another. Imagine how a patient might feel when he or she is interacting with a health-care provider who does not make meaningful eye contact, does not smile, and has conversations with their back turned. Not good.

If you have frequent face-to-face contact with patients and you fail to make meaningful eye contact with them, rarely smile, and turn away from them during conversations don't be surprised if they are making ugly faces at you and using vulgar hand gestures while your back is turned.

There are many more ways to nonverbally communicate with patients but let's talk about one more that has a tremendous positive impact on both the patient and health-care provider. Let's talk about the power of touch.

Touch has long been considered a sign of positive emotion, comfort, and compassion.[31] Among family members the amount of touching, such as embraces and arms around shoulders, is positively associated with the intensity of smiling amongst family members and is a predictor of similar positive behaviors outside of the home as well.[32]

A gesture of lightly touching a person, such as on the shoulder, communicates warmth and comfort and is reported to have a positive effect on the feelings of the person being touched.[33]

In 2011 researchers provided a neurophysiologic rationale for the effects of touch on patient's perceptions and the level of trust the patient placed in the health-care provider. They found that the hormone *oxytocin* was released in the brain of the patient as a result of receiving a comforting touch from the health-care provider and also the patient's perception of trust between themselves and the provider.

The release of oxytocin in the patient's brain caused the patient to

---

31   Matsumoto, *Nonverbal Communication: Science and Applications*, 86.

32   Christopher Oveis, June Gruber, Dacher Keltner, Juliet. L. Stamper, and W. Thomas Boyce, "Smile Intensity and Warm Touch as Thin Slices of Child and Family Affective Style," *Emotion* 9, no. 4 (August 2009): 544–548.

33   Demarais, *First Impressions: What You Don't Know about How Others See You*, 171.

report feelings of well-being, a heightened pain threshold, and many other positive physical and mental states.[34]

Nothing is more powerful than taking the hand of a patient or lightly touching the person on the shoulder when he or she is emotionally vulnerable and struggling through a difficult medical situation.

Touch is a basic human need. Without touch humans often fail to bond and form meaningful attachment to others. Touch is a one of our primary senses and a powerful form of communication. Touch is a part of how we sense, evaluate, and describe the world around us.[35]

Touch connects the health-care provider and patient in ways words never can. Touching the patient should be a strong reminder to any health-care provider that they are caring for a human being and not a disease or unit of procedural volume.

Now that we have covered some of the basic yet powerful nonverbal behaviors to be exhibited during the patient encounter, let's move on to verbal behaviors that demonstrate caring and reinforce to the patient they are within a caring culture.

---

34   Michael M. Patterson, "Touch: Vital to Patient-Physician Relationships," *The Journal of the American Osteopathic Association*, 112, No. 8 (August 2012): 485.

35   Mitchell L. Elkiss, and John A. Jerome, "Touch: More Than a Basic Science," *The Journal of the American Osteopathic Association* 112, no. 8 (August 2012): 514–517.

CHAPTER 4

Verbal Behaviors of a Caring Culture

There is a reason we covered nonverbal behaviors first. In many ways nonverbal behaviors are more influential in creating the patient's perception of a caring culture. "Actions speak louder than words," is more than just a cliché. Patients see us before they hear us. And when we do get around to speaking they watch us closely to see if our nonverbal behaviors support the words coming out of our mouths.

Even though this writer places great emphasis on nonverbal behavior the power of purposeful spoken words cannot be underestimated. The reason so many health-care providers use the communication techniques of AIDET[36], RELATE[37], and patient-centered explanations with positive intent[38] is because they work.

Staff training, however, often places great emphasis on verbal communication techniques with little or no attention paid to nonverbal communication. That approach is ill advised and can be likened to putting the cart before the horse.

Depending upon the culture of the group or organization, providing a script for staff to follow when interacting with patients could prove to

---

36 Quint Studer, *Hardwiring Excellence* (Gulf Breeze, FL: Fire Starter Publishing, 2003), 94.

37 "RELATE," Baptist Health Care Leadership Group, accessed November 14, 2013, http://www.bhclg.com/relate-online-courseware.

38 Leebov, *Wendy Leebov's Essentials for Great Patient Experiences: No-Nonsense solutions with Gratifying Results*, 17—36.

be ineffective if staff generally avoid eye contact when passing people in a hallway, do not verbally acknowledge others in the hallway or elevators, and seem to rarely smile.

If you have done staff training on scripting one simple way to gauge the effectiveness of that training is to observe a staff member interacting with a patient. If the staff member is looking away or has his or her back turned while following a script then rest assured your training with that individual has come up short.

There are many different communication techniques being promoted within the health-care industry. Regardless of the tool's acronym the purpose of tool is the promotion of meaningful conversation. Improving the patient's overall experience, helping the patient to better understand his or her current state of health, clarifying how the provider is going to serve the patient's health needs, and helping the patient interpret the provider's intentions are the *why's* behind purposeful communication.

Most health-care providers approach the patient encounter with the best of intentions. Unfortunately patients cannot always interpret the provider's intentions. Rather than take an unnecessary risk and leave the interpretation to chance the provider must be proactive in helping the patient understand the provider's role in the patient-provider relationship.

Research shows a strong positive relationship between a health-care provider's communication skills and a patient's capacity to follow through with medical recommendations, self-manage a chronic condition, and adopt preventative health behaviors.[39] In 2009 a meta-analysis of more than a hundred empirical studies found that provider communication is significantly positively correlated with patient adherence to medical directives.[40]

Before we examine the use of specific words and phrases let's take a look at the components of compassionate communication. Andrew Newberg, MD, and Mark Robert Waldman have identified twelve components of

---

[39] "Impact of Communication in Healthcare," Institute for Healthcare Communication, accessed September 23, 2013, http://healthcarecomm.org/about-us/impact-of-communication-in-healthcare.

[40] Kelly B. Haskard Zolnierek, and M. Robin DiMatteo, "Physician Communication and Patient Adherence to Treatment: A Meta-analysis," *Medical Care* 47, no. 8 (August 2009): 826–834.

compassionate communication in their powerful book *Words Can Change Your Brain*.[41]

The components include six steps to take (preparatory) before one engages another person in a conversation.

1. Relax
2. Stay present
3. Cultivate inner silence
4. Increase positivity
5. Reflect on your deepest values
6. Access a pleasant memory

Once the interaction begins the seventh step becomes what Newberg and Waldman call the most crucial aspects of communication, and that is:

7. Observe nonverbal cues

When the conversation begins, the last five strategies one should consistently adhere to include the following:

8. Express appreciation
9. Speak warmly
10. Speak slowly
11. Speak briefly
12. Listen deeply

The concept of compassionate communication is powerful and you are strongly encouraged to read Newberg and Waldman's book in order to gain a richer appreciation for the science, strategies, and practical applications of compassionate communication. If you and your teams are using scripts everyone will prove to be more effective if the delivery of the script is aligned with caring and compassion.

---

41   Andrew Newberg, and Mark Robert Waldman, *Words Can Change Your Brain: 12 Conversation Strategies to Build Trust, Resolve Conflict, and Increase Intimacy*, (New York, NY: Penguin Group, 2012) 121—124.

That brings us right back to the functionality of the left and right hemi-spheres of the brain and how that plays out during the patient encounter. It is also the reason why nonverbal communication takes on a significant level of importance during patient interactions. The left brain specializes in *text* and the right brain specializes in *context*.[42]

The left brain focuses on what is said and the right brain focuses on how it is said. As we learned in an earlier chapter the patient equates the quality of their clinical care to the quantity of caring delivered during their health-care encounter. If we want to create an incredible patient experience we had better learn to connect with the patient's right brain expectations.

Specific scripts are not going to be provided in this book. We will leave that up to the experts who already have supplied us with an endless list of useful and powerful phrases and the framework in which the phrases are to be delivered. We are, however, going to focus on one script all providers need to be comfortable using during the patient encounter. In the next chapter we will review the importance of responding to the concerns and complaints of patients and the language we need to use in those situations to build the perception of advocacy.

---

42   Daniel H. Pink, *A Whole New Mind: Moving from the Information Age to the Conceptual Age*, (New York, NY: Penguin Group, 2005) 20.

CHAPTER 5

Responding to Patient Complaints

Because of a multitude of factors it is not uncommon for a patient or the patient's family to complain about something they experienced during their health-care journey. The health-care provider's response on the front-end to the patient's complaint influences the likelihood of successful complaint resolution on the back end and also strongly influences the patient's overall satisfaction. Let's talk more about complaint response and complaint resolution.

The response is what happens in the first thirty or sixty seconds after the patient delivers the complaint to the health-care provider. The response focuses on the immediate nonverbal and verbal behaviors of the health-care provider that receives the complaint.

The resolution, sometimes called the service recovery, is what unfolds following the complaint and response. The resolution process usually involves an investigation, creation of corrective measures, and communicating with the patient several days or weeks following the complaint.

Our focus will be on delivering the proper response and why that is critically important to the patient's overall satisfaction. It is the response that sets the stage for successful resolution of the complaint. If the initial response is received well the patient then considers the health-care provider to be an ally. If the patient views the provider as an ally the patient often times will be more willing to work with the provider toward a satisfactory resolution.

Let's review the three basic steps and scripts (text) of successfully responding to a patient complaint and the framework (context) in which the response should be delivered.

Immediately after the complaint is received, the first script is the apology. The immediate verbal response should be similar to, "I am so sorry that happened," or "I am sorry you had that experience."

When the apology is immediately delivered by the individual receiving the complaint that person must be facing the patient, making direct eye contact, and displaying a facial expression of interest and concern. Both text and context must complement each other and be appropriate for the situation at hand. There can be no mixed messages here.

The apology is how we as human beings communicate compassion, empathy, and caring to those around us. Here is an example of the power of the apology. Think about a time when someone close to you shared with you something very unfortunate that happened in his or her life. Maybe that person lost a loved one or a favorite pet passed away. As the individual told you his or her story more than likely tears came to their eyes which brought tears to your eyes as well.

What did you say and do in response to the story they shared? Chances are you were facing them while they told you their story. You probably made and sustained direct eye contact with them as you listened intently. You nonverbally expressed concern, compassion, and caring through your facial expressions. At least once you probably said, "I am so sorry." All of those actions are how we naturally demonstrate caring, compassion, and concern to those around us.

For readers worried about the implications an apology may have in a patient complaint or grievance situation rest assured the apology is often not considered an admission of guilt. In thirty-six states an apology offered purely as an expression of regret that something occurred is statutorily protected and not an admission of liability.[43] For more information about the power of the apology visit the web site *Sorry Works!* www.sorryworks.net.[44]

Immediately after the apology we need to thank the patient for speaking up and bringing his or her complaint to our attention. Many patients are reluctant to speak up when dissatisfied. They keep it to themselves and choose to suffer in silence.

But why does that happen? What holds a patient back in expressing his

---

43 David A. Jones, "Apology Laws Foster Compassion," *Provider,* (May 21, 2012), accessed online September 24, 2013, http://www.providermagazine.com/columns/Pages/Apology-Laws-Foster-Compassion.aspx.

44 Sorry Works! Accessed online September 24, 2013. http://www.sorryworks.net/.

or her displeasure? What does he or she fear? The patient fears retaliation and retribution from his or her health-care providers for complaining. The sad reality is that health-care providers, when they begin to lose emotional control in stressful situations, do retaliate against patients and the patient's family.

Let's be crystal clear about the type of retaliation that takes place. We are not talking about beating a patient with a rubber hose or purposely causing harm to the patient. The most common retaliation is the way a patient is treated verbally and nonverbally.

When health-care providers get frustrated and lose emotional control retaliatory behaviors follow. Typical retaliatory behaviors include making less eye contact with the patient, smiling less frequently in front of the patient, and not facing the patient during conversations. Sound familiar?

Remember the exercise we went through in a previous chapter and the two people you were asked to picture in your mind? That is just one example of how a provider retaliates against a challenging patient or family. Rolling of the eyes, deep sighs, and facial expressions that communicate annoyance are common retaliations health-care providers resort to when frustrated with the situation and loosing emotional control.

Fear of retaliation is the reason many of us are reluctant to send our less-than-satisfactory meal back to the kitchen in a restaurant. It's because in the back of our mind we are fearful about what the cook might do to our food just before the waiter brings it back to us. That fear keeps us from complaining so we eat slower, pick at our food, and wait for dessert.

The whole time we are bad-mouthing the restaurant to those with us and we can't wait to post or text our experience on our favorite social media site. Patients and families are the same way about their health-care experience.

The third and final script in responding to a patient complaint is to make a commitment that action will be taken. The commitment-to-action script is, "Now that you've made me aware of your concern I can begin to address the issue."

The commitment to action does not identify a specific solution to the complaint. The commitment to action is a confirmation that the patient's voice has been heard and action will be taken on the complaint.

The resolution process, which follows the response, is dependent upon the department or organization's policies and procedures. The response sets the stage for a successful resolution. The response is critically important to the patient as it builds the perception of advocacy and establishes the health-care provider as an ally addressing the complaint and working on behalf of the patient.

The health-care provider's immediate response to a patient concern or complaint is critically important to patients across many settings. Press Ganey is one of the largest health-care consulting firms in the United States that specialize in patient satisfaction survey research and quality improvement services. The service Press Ganey provides includes satisfaction survey development, printing, distribution, collection, analysis, and reporting.

In 2011 Press Ganey published the *Pulse Report: Perspectives on American Health Care* which described the experiences of patients in many settings including both inpatients and outpatients.[45] Within the report Press Ganey provides a national priority index which is a prioritized list of survey questions that statistically demonstrate strong correlation to the patient's overall satisfaction and loyalty and question performance based upon the average reported mean score.

The priority index reflects service issues that are of high importance to the patient and requires prioritized focus improvement if mean score performance on the question is low.

The specific question, *Staff's response to concerns or complaints during my stay/visit,* is found on both the Press Ganey inpatient and outpatient satisfaction survey. The question was reported by Press Ganey to be number one on the inpatient nationwide priority index based upon more than 2.8 million patient responses. The question was also the number one outpatient nationwide priority index item based upon more than 2.3 million patient responses.

Interestingly enough, *response to concerns and complaints* was ranked number one on the inpatient and outpatient 2010 and 2009 Press Ganey nationwide priority index as well.[46] Is how we respond to a patient's concerns and complaints important? You bet it is!

Effectively responding to the patient's complaints demonstrates advocacy which is critically important since patients often feel victimized by the service delivery processes health-care systems create.

---

45 Press Ganey, *2011 Pulse Report: Perspectives on American Health Care*, (South Bend, IN: 2011), accessed online September 29, 2013. http://www.pressganey.com/researchResources/hospitals/pulseReports.aspx.

46 Press Ganey, *2011 Pulse Report: Perspectives on American Health Care*, http://www.pressganey.com/researchResources/hospitals/pulseReports.aspx.

In Chapter 1 we reviewed how left-brain-oriented service delivery and right-brain-oriented service receipt creates conflict for the patient during their health-care experience. That conflict inherently contributes to patient dissatisfaction because the patient often feels victimized by the service delivery process. When people feel victimized they need and desire advocates to work on their behalf. *Response to concerns and complaints* is one of the best ways a health-care provider can demonstrate advocacy and caring during the patient encounter.

In the next chapter we will review a strategy called patient experience mapping that helps identify critical touch points along the patient's journey where specific verbal and nonverbal behaviors are best put to use.

CHAPTER 6

The Patient Experience

The sum of all interactions, shaped by an organization's culture,
that influence patient perceptions across the continuum of care.
-The Beryl Institute

The very essence of the patient experience is neatly summarized by The Beryl Institute[47] in the above definition. The patient experience is the sum of all interactions the patient has during their journey. All of the interactions between the patient and the organization is shaped by the organization's culture. Both culture and interactions influence the patient's perceptions throughout the patient's continuum of care.

If our objective is to exceed patient expectations and achieve service excellence we must first understand exactly what the patient experiences during their journey. We cannot hope to manage the patient experience if we do not have intimate knowledge of all touch points and moments of truth the patient encounters while they are with us.

To fully understand the patient's journey, the entire continuum of the patient's care, we must first break it down and map out each step of the encounter. By identifying each step of the patient's journey we gain a richer understanding and greater appreciation of all the moments value is added to the patient experience and the moments it is not.

---

47 "The Patient Experience," The Beryl Institute, accessed November 15, 2013, http://www.theberylinstitute.org/?page=DefiningPatientExp

The method often used to break down and map out the patient's journey is called patient experience mapping. It is similar to creating a Lean Six Sigma[48] value stream map where the steps of a process are identified and defined to ensure the right tools are being used at the right times in the right ways. Lean Six Sigma value stream mapping allows us to identify and remove unnecessary steps which improves performance and adds value to the process.

Patient experience mapping is similar in that the patient's journey is diagramed and defined to help ensure everyone interacting with the patient is using the right tools at the right times in the right way. Our objective is to find ways to add value to the patient's experience while service is delivered.

Only a high-level description of the patient experience mapping process will be provided in this book. The actual process is lengthy, time consuming, and extremely detailed. Even though the amount of time and resources an organization must dedicated to the experience mapping process is significant, the value gleamed from the experience mapping process will prove to be invaluable in improving the satisfaction of patients being served.

To start, a team must be assembled to create the patient experience map. The team should be comprised of individuals representing the multiple touch points the patient will encounter such as registration, lab, medical imaging, perioperative services, and so on. Over time more people may be brought onto the team and some members will leave depending upon the team's needs. Having multiple eyes and staff perspectives to help break-down, define, and diagram the patient journey will pay big dividends in the long run.

Once the team is assembled the patient experience mapping begins by visually identifying the multiple steps a patient encounters receiving a particular service. It begins by team members listing all the interactions and events between the patient and the organization during the journey such as finding a parking space, walking to the facility entrance, entering the building, presenting at registration, and so on all the way to the end when the patient exits the building, returns to their car and leaves the campus.

Each and every step, from beginning to end is clearly identified and defined. All of the details of the patient experience is mapped out to give a

---

48   John Morgan and Martin *Brenig*-Jones, *Lean Six Sigma For Dummies*, (Chichester, West Sussex, England: John Wiley & Sons, Ltd., 2009).

visual representation of what the average patient experiences during their time with the organization.

At each touch point when there is interaction between staff and patient, the actions and activities of both staff and patient must be clearly described and defined. For example, at registration this might include entering demographic data into the electronic medical record, scanning insurance cards, or signing documents.

Between each touch point we need to identify the time spent when value is not being added to the patient experience. Examples of non-value added moments would be the time the patient spends sitting in a waiting room, changing clothes in a dressing room, or walking in a corridor between touch points.

The actual mapping process uses shapes to represent common types of events of the patient's experience and written descriptions of each step of the process.

For example, a simple patient experience map may use the oval shape ⬭ to identify non-value added time, or the plus shape ⊹ to identify moments where the patient interacts directly with staff and values is added. The arrow shape ⬜⟩ can be used to represent moments of down-time when the patient is moving between touch points such as walking down a corridor.

Steps in a simple patient experience map may look like this:

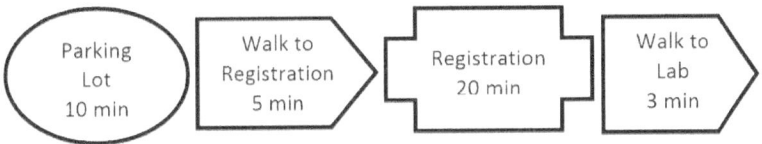

The mapping team will work to bring details to the map and help ensure all aspects of the patient experience is inventoried and documented.

Once the initial patient experience map is created it is important that members of the mapping team actually participate and walk through the diagramed patient journey to ensure an accurate picture of the patient experience has been captured. This activity also provides team members a prime opportunity to shadow and interview actual patients going through the journey real-time.

Patients are often eager to disclose their expectations and perceptions when asked and many will never shy away from making suggestions for improvement. Team members who shadow and interview patients should provide their business card or contact information should the patient wish to give additional feedback after their experience ends.

Team members must actually walk through each and every step of the patient experience just as the patient would. If you want to know what the patient experiences finding a parking spot in the parking garage then a team member needs to drive their car into the patient parking garage and find a place to park.

If the team wants to know what it is like for a patient to sit in a dressing room for 15 minutes wearing a hospital gown waiting for a procedure to begin then a team member needs to undress, put on a hospital gown and sit in a dressing room for 15 minutes. Members of the experience mapping team will probably be amazed at what they hear while sitting in a patient dressing room.

Obviously there will be some limitations as to what the team member will experience in these efforts. No one expects a team member to have their blood drawn needlessly just to have a true patient experience. Secret shoppers, real patients having needed tests and treatments, can provide immediate feedback regarding their experience for that particular purpose.

Secret shoppers are valuable resources available to gain insight into the patient experience. A secret shopper can be selected before the day of service or they can be contacted after the service through a post-visit phone call. Their feedback will prove to be invaluable.

Secret shoppers can also be employees having tests or treatments. Many organizations provide some type of reporting mechanism for employees to provide feedback regarding their experience as a patient within the organization. The voice of the patient/employee can be captured in numerous ways and it will prove to be incredibly valuable in the patient experience mapping effort.

Sometimes we lose sight of the fact that employees and physicians are also consumers of health-care services and their voice is just as valuable as nonemployees and health-care providers.

Now that the map has been created the team begins to identify the right tools for staff to use at the right times and in the right ways at critical touch points. The team creates verbal and nonverbal instructions for staff to use in helping shape the patient's expectations, building perceptions, and proactively preparing the patient for future steps in their journey.

Many times when scripting is created it is focused on a single moment in time at a single touch point. Unfortunately there are opportunities lost in taking this silo approach to handling the patient experience. The experience mapping exercise provides the mapping team a view of the entire continuum of care. This enables the team to develop tools for staff to use at a given moment of time and tools for staff to use that prepares the patient for future experiences in their journey.

For example, the verbal communication tools staff could be given might be, "Good morning Mr. Johnson. My name is Bill and it's a pleasure being with you today. It will only take me about 5 minutes to verify the information we have on file for you in the medical record is correct. Once we have that taken care of I will be happy to escort you to the outpatient testing area and introduce you to the care team in the lab. They will take excellent care of you from there." This simple script serves the need of patient and employee at the given moment but also prepares the patient for the next step in their journey.

Each step, each touch point is prepared in this manner. The patient experience mapping exercise also provides the opportunity to find and eliminate steps proven to be wasteful and non-value adding. It also provides opportunity to create and add value during perceived non-value added moments. This is where it is helpful to have staff from marketing or public relations on the mapping team. These individuals will help find ways to bring valuable information to the patient during down-time and make non-value added moments more valuable.

For example, waiting rooms are often non-value added moments where patient satisfaction frequently declines. Many times the solution is to put televisions and newspapers in a waiting room in the hope that entertaining the patient during down time reduces frustration with waiting and results in a happier patient. Having members of the marketing and public relations department on the mapping team brings a different set of eyes and skill sets to the improvement process. The waiting room is also a great place for members of the mapping team to sit and interview patients regarding their perceptions of what brings true value to the waiting experience.

So now how do we bring all of this together and make it work? How do we ensure members of the health-care team actually use the tools we give them and create a caring culture for the patient, visitor, family member, and other co-workers? In the next chapter we will explore the steps leaders need to take to ensure a caring culture is created and sustained.

CHAPTER 7

Staff Engagement and Accountability

ringing all the information provided in the previous chapters together and making a caring culture reality is challenging. This entire book could be dedicated to how cultures are created and sustained. Given this book's limitations, a high-level overview will be taken of staff training, coaching, and reinforcing of caring behaviors and the staff's execution of caring initiatives. Changing a department or organizational culture is tough and time-consuming.

First let's talk about staff training. Most health-care organizations in the United States do a fine job training staff on new initiatives and performance expectations. In most health-care organizations a tremendous amount of resources is spent annually informing and educating staff on what they need to know to get the job done and done right. There are many ways hospitals inform and educate staff, including class lectures, staff meetings, videos, and computer-based learning (CBL) modules. Most training methods are effective at disseminating information to the masses. The problem lies in the effectiveness of training.

Mike Easterday discusses staff training and changing staff behavior in an article in *The Satisfaction Monitor* entitled, "Our People Were Trained So Why Don't They Act like It?"[49] Mr. Easterday described standard training methodologies as the old sheep dip philosophy of sending people to a class and then expecting new behaviors as a result. That rarely happens.

---

49   Mike Easterday, "Our People Were Trained … So Why Don't They Act Like It," *The Satisfaction Monitor*, (South Bend, IN: Press Ganey, 2000), 1—2.

Easterday points out that certain factors increase training effectiveness, including the following:

1. Leadership must utilize a structured reinforcement process to help develop new behaviors. This would include leader observation, feedback, and coaching.
2. Management must also be held accountable by senior leadership to understand, support, role model, and coach the new behaviors.
3. Clear service-based and principle-based behaviors [caring behaviors] must be clearly communicated and explained to all employees. Employees must believe in the principles behind the new proposed behaviors to incorporate them into their own values, attitudes, and behaviors.

Training, Easterday points out, must reach the employee on three dimensions—intellectually, emotionally, and creatively. The intellectual dimension is the logical part where we think, learn facts, and collect information [left brain]. The emotional dimension is the feeling part [right brain] of the training, which is important because the employee can't just *know* what to do, they must also *feel* like doing it.

The deepest dimension, the creative dimension, is the level that houses our values and beliefs. It is this internal unconscious dimension that drives our feelings and ultimately our actions. To effectively train staff we must be able to reach staff on all three dimensions in order to achieve behavioral change.

Some of the ways to reach all three dimensions when training staff include the following:

1. Teach staff how to emotionally connect with both their head and their heart.
2. Ensure staff understands that service excellence involves more than words. It includes a genuine smile, a comforting touch, or taking a moment to demonstrate true compassion and empathy.
3. Help caregivers understand the true purpose of their job is delivering patient-centered care and caring and how that purpose directly relates back to the departments and organization's performance and success.

Effective training is still not enough to achieve behavioral change. We must be sure staff explicitly understand what is expected of them. We will not be able to hold staff accountable for what they do not completely understand. If the staff cannot speak to expected behaviors they will not be able to perform them. That is the value of auditing staff knowledge of performance expectations. The auditing process is similar to an interview. A leader asks each staff member to verbally describe performance expectations and then explain how the staff member will bring those expectations to life when interacting with a patient.

For example, if a staff member was trained to initiate the patient encounter with three nonverbal behaviors such as maintaining eye contact, smiling, and facing the patient while speaking, the audit question would be, "Please tell me the three nonverbal behaviors you will use at the beginning of the patient encounter." The expectation is the staff member will answer, "I will maintain eye contact, smile, and face the patient when I speak."

Auditing staff's knowledge of performance expectations is important because it helps ensure every staff member is on the same behavior standards page of the caring culture handbook. Remember, staff will not exhibit the expected behaviors if do not know them. If they do not know them they cannot speak to them. After training is completed leadership must take time to perform a knowledge audit to ensure staff's readiness to move on to the next step.

Once the training and knowledge audit steps are completed the leader must personally demonstrate and model the expected behaviors staff are expected to emulate. Leadership does not have the luxury of telling staff, "Do as I say, not as I do."

To achieve behavioral change training is not enough. Simply testing staff knowledge post training will not ensure compliance and behavior modification. Staff must be shown, by leadership, what to do and what to say. The new caring behaviors must be demonstrated through leader role modeling so that staff members explicitly understand what is expected of them.

Leader role modeling is more than just putting your money where your mouth is. When the leader models the expected caring behaviors it not only provides staff with an example of exactly how the behaviors should be performed but also enables the leader to use one of the most powerful and influential training tools available—the vicarious experience.[50]

50   Kerry Patterson, Joseph Grenny, David Maxfield, Ron McMillan, and Al Switzler, *Influencer: The Power to Change Anything*, (New York, NY: McGraw-Hill, 2008) 53—57.

When we are learning to do something new for the first time most of us are shown how to do it correctly before we go it alone. The vicarious experience gives the learner a real-time example of what is to be done before the new task is attempted.

Once the training is completed, the knowledge audited, and expectations modeled it is time for the leader to ensure staff are performing the new behaviors and meeting expectations. The leader must adopt, as Easterday puts it, a structured reinforcement method to ensure staff are executing the caring culture strategic plan.

The structured reinforcement Easterday references is nothing new. It has been a management tool for decades. Over the years it has gone by many different names, one of which is management by wandering around (MBWA), introduced to us by Tom Peters and Robert H. Waterman Jr., in their 1982 best seller *In Search of Excellence: Lessons from America's Best-Run Companies.*[51]

William Hewlett and David Packard, founders of Hewlett Packard (HP), effectively used this management tool within HP in the 1970's. It allowed leaders to randomly mingle with staff in the workplace and observe worker performance, coach and council staff, give staff feedback on processes, and make operational adjustments as needed.

A similar tool, rounding for outcomes, is a method advocated by Quint Studer.[52] Compared to MBWA, rounding for outcomes is a more structured practice by a leader to observe and interact with staff, reinforce performance expectations, coach and council as necessary, and achieve staff engagement.

These leadership techniques are extremely useful and allow leaders to keep their fingers on the pulse of staff behavior and performance. Through observation, (standing back while watching and listening), shadowing (closely following staff and observing their performance in real time), and secret shopping (actual patients receiving service and giving firsthand evaluation feedback of staff performance), the leader has the ability to reinforce service excellence expectations and hold staff members accountable for their performance.

---

51 Thomas J. Peters, and Robert H. Waterman, Jr., *In Search of Excellence: Lessons from America's Best-Run Companies*, (New York, NY: Warner Books, 1982) 122.

52 Quint Studer, *Results that Last: Hardwiring Behaviors that Will Take Your Company to the Top*, (Hoboken, New Jersey: John Wiley & Sons, Inc., 2008) 25—33.

These steps help ensure that behaviors change over time and that a new behavioral status quo is developed within the work culture. The process never stops. It is an ongoing endeavor compared to running a marathon that has no finish line. Leaders can never turn their backs on the culture being reshaped. Remember, if the leader does not manage the culture, the culture will eventually manage the leader. The following change process must be completed full circle and then repeated to ensure cultural change takes place and becomes permanent.

The first time an employee is trained, the steps should be followed in the above order. Once the employee has experienced the above process full circle it is not necessary to start again with effective training and go through the entire process a second time. Sometimes it is a matter of role modeling more than once. Sometimes it is necessary to do extra observation, shadowing, and coaching. Each employee is different and reshaping behaviors will vary depending upon the needs and readiness of the employee.

So what do we do with employees that just don't get it? What do we do with employees that have gone through the above change process several times and their behavior does not change? Unfortunately their lack of performance is holding culture change back and, worse of all, they are a negative influence on other staff members. How do we as leaders decide when to continue coaching and role modeling a staff member and when to administratively address noncompliant behavior?

Every leader must choose the course to take with an employee whose performance does not meet expectations. Every organization has their own policies and procedures regarding corrective action and possible termination. The leader must abide by those processes within their respective organizations.

The decision on when to provide remedial training, administer corrective action, or initiate the termination process is driven by what takes place in step-four of the change process.

After a full change process cycle has been completed with an employee the leader should expect to see new behaviors being exhibited. The new behaviors may not be perfect. It takes time to get new behaviors right. Frequently going over steps two, three, four, and five of the change process will help refine the employee's new behaviors.

If an employee has gone through the full cycle of the change process and has repeated steps two, three, four, and five and new behaviors are still not being exhibited, then the leader is more than likely facing a compliance issue and insubordinate behavior from the staff member.

The leader must make absolutely sure all five steps of the change process have been completed and repeated to ensure every effort has been taken to help the employee be successful in changing their behavior. If, after all the education and training, there is still no behavioral change the leader must then ask, "Is this employee simply refusing to comply with the new behavioral expectations?"

If the answer is yes then there is high culpability on the part of the employee and the leader must address it aggressively. Knowing what to do and purposely not doing it is insubordination and must be dealt with immediately and aggressively.

When it comes to behavior culpability for one's actions increases as purposeful behavior increases. As culpability increases so do the positive and negative consequences of the purposeful behavior. If health-care providers do not exhibit the expected caring behaviors simply because they are confused or truly don't explicitly understand what is expected of them then there is low culpability. However, if the health-care provider can speak to the expectations (step two) and the expectations have been adequately modeled by the leader (step three) and the employee simply will refuses to perform the expected behaviors (step four), then culpability is high and worthy of extreme consequences.

The more purposeful the destructive behavior, insubordination and noncompliance with directives is considered to be extremely destructive, the more culpable the individual is for their decisions and consequently the more severe the consequences the individual should face.[53]

Ultimately all the work and resources being used to reshape the existing culture into a more caring culture is more than justified for the benefits provided the patient. Creating and sustaining caring cultures ensures patient expectations will consistently be exceeded, drives patient satisfaction, and has a positive influence on the patient's clinical outcomes.

We have covered a lot of information in the previous chapters. The reader should at this point be better prepared to begin reshaping the existing culture into a more caring culture. Once the culture has been changed the leader can never turn his or her back on the new culture. Cultures must be continually maintained, nurtured, and groomed if they are to thrive and flourish.

In the final chapter we talk about the importance of leadership in cultural change. Ultimately it is the quality of leadership that makes or breaks the ability to successfully create and sustain a caring culture.

---

53   James Reason, *Managing the Risks of Organizational Accidents*, (Burlington, VT: Ashgate Publishing Company, 1997), 209.

CHAPTER 8

It All Comes Down to Leadership

If your actions inspire others to dream more, learn more,
do more and become more, you are a leader.
—John Quincy Adams, 1767–1848
Sixth President of the United States of America

Leadership is required to change a culture. Not just any kind of leadership. It must be purposeful leadership. It must be intentional leadership.[54] It must be leadership that drives results, never allowing chance or luck to enter into the outcomes equation.[55] Leadership must be capable of and courageous enough to change an existing culture and maintain it over time.

A leader must make a conscious decision and put forth the effort to become the most effective leader he or she can be.[56] That is what being a leader is all about and that is what it takes to create a caring culture in the workplace. Changing the work culture requires a leader who can persevere in the long run and is willing to make personal sacrifices to achieve the goal.

---

54 Kenneth A. Shaw, *The Intentional Leader*, (Syracuse, NY: Syracuse University Press, 2005) 5—20.

55 Joe Calloway, Chuck Feltz, and Kris Young, *Never by Chance: Aligning People and Strategy Through Intentional Leadership*, (Hoboken, New Jersey: John Wiley & Sons, Inc., 2010) 131—143.

56 Eric Papp, *Leadership by Choice: Increasing Influence & Effectiveness through Self-Management*, (Hoboken, New Jersey: John Wiley & Sons, Inc., 2012) xvi.

If you are in any type of leadership position, formal or informal, and have read this book you are now in a position to begin changing a culture and improve what your patients experience and improve the service they receive. It is entirely up to you the path you choose. To reshape the existing culture into a more caring culture you will need to choose the path least taken. Which path is that? The uphill and rocky path most people avoid.

You can make a difference. You need to be the one to make the difference. The lives of your patients, their families, the physicians caring for those patients and all your coworkers will become better if you choose to make a difference. No matter what level of leadership you hold, formal or informal, you have the power to make a difference. You simply need to step up, be a true leader and be the catalyst of change. The only prerequisite is that you care a whole awful lot for what matters most.

> *Unless someone like you cares a whole awful lot,*
> *Nothing is going to get better. It's not.*[57]

---

57   Dr. Seuss, [Theodor Seuss Geisel], *The Lorax,* (New York: Random House, Inc., 1971), 62.

BIBLIOGRAPHY

Ambadar, Zara, Jeffrey F. Cohn, and Lawrence I. Reed. "All Smiles are Not Created Equal: Morphology and Timing of Smiles Perceived as Amused, Polite and Embarrassed/Nervous." *Journal of Nonverbal Behavior* 33, no. 1 (March 1, 2009): 17–34.

Baker, Susan Keane. *Managing Patient Expectations: The Art of Finding and Keeping Loyal Patients.* San Francisco, CA: John Wiley & Sons, 1998.

Baptist Health Care Leadership Group. "RELATE." Accessed November 14, 2013. http://www.bhclg.com/relate-online-courseware.

Bartlett, Edward E., Marsha Grayson, Randol Barker, David M. Levine, Archie Golden, and Sam Liffer. "The Effects of Physician Communications Skills on Patient Satisfaction; Recall, and Adherence." *Journal of Chronic Diseases* 37, no. 9–10 (1984): 755–64.

Boudreaux, Edwin D., and Erin. L. O'Hea. "Patient satisfaction in the Emergency Department: A review of the literature and implications for practice." *The Journal of Emergency Medicine* 26, no. 1 (2004): 13–26.

Boulding, William, Seth W. Glickman, Matthew. P. Manary, Kevin A. Schulman, and Richard Staelin. "Relationship between Patient Satisfaction with Inpatient Care and Hospital Readmission within 30 Days." *American Journal of Managed Care* 17(1) (2011): 41–48.

Calloway, Joe, Chuck Feltz, and Kris Young. *Never by Chance: Aligning People and Strategy Through Intentional Leadership.* Hoboken, New Jersey: John Wiley & Sons, Inc., 2010.

Demarais, Ann, and Valerie White. *First Impressions: What You Don't Know about How Others See You.* New York, NY: Bantam Books, 2004.

Easterday, Mike. "Our People Were Trained … So Why Don't They Act Like It." *The Satisfaction Monitor.* South Bend, IN: Press Ganey, 2000.

Elkiss, Mitchell L., and John A. Jerome, "Touch: More Than a Basic Science." *The Journal of the American Osteopathic Association* 112, no. 8 (August 2012): 514–517.

Engel, Marcus. The *Other End of the Stethoscope: 33 Insights for Excellent Patient Care.* Orlando, FL: Ella Press, 2006.

Galanti, Geri-Ann Galanti. *Cultural Sensitivity: A Pocket Guide for Health Care Professionals,* 2nd ed. Oakbrook Terrace, Illinois: Joint Commission Resources, 2012. http://store.jcrinc.com/cultural-sensitivity-a-pocket-guide-for-health-care-professionals-second-edition.

Garman, Andrew N., Joanne Garcia, and Marcia Hargreaves. "Patient Satisfaction as a Predictor of Return-to-Provider Behaviors: Analysis and Assessment of Financial Implications." *Quality Management in Health Care* 13, no. 1 (Jan.–March 2004): 75–80.

Heskett, James L., Thomas O. Jones, Gary W. Loveman, W. Earl Sasser, Jr., Leonard A Schlesinger. "Putting the Service-Profit Chain to Work." *Harvard Business Review* 72, no. 2 (March–April 1994): 164–174.

Hirsch, Lonnie. "15 Best Practice Reasons Professionals Care about Patient Satisfaction," *PatientExperience.com.* http://patientexperience.com/15-practice-reasons-professionals-care-patient-satisfaction/.

Institute for Healthcare Communication. "Impact of Communication in Healthcare." Accessed September 23, 2013. http://healthcarecomm.org/about-us/impact-of-communication-in-healthcare.

Jones, David A. "Apology Laws Foster Compassion." *Provider*. (May 21, 2012), accessed online September 24, 2013, http://www.providermagazine.com/columns/Pages/Apology-Laws-Foster-Compassion.aspx.

Jones, Susan. "Nature and Nurture in the Development of Social Smiling." *Philosophical Psychology* 21, no. 3 (June 2008): 349–357.

Kelly, Bob. *Worth Repeating: More Than 5,000 Classic and Contemporary Quotes*. Grand Rapids: Kregel Publications, 2003.

Kipp, Kris M. "Implementing Nursing Caring Standards in the Emergency Department." *The Journal of Nursing Administration* 31, no. 2 (2001): 85–90.

Kohlrieser, George. *Hostage at the Table: How Leaders Can Overcome Conflict, Influence Others, and Raise Performance*. San Francisco, CA: Jossey-Bass, 2006.

Lee, Fred. *If Disney Ran Your Hospital 9 ½ Things You Would Do Differently*. Bozeman, MT: Second River Healthcare Press, 2004.

Leebov, Wendy. *Essentials for Great Patient Experiences: No-Nonsense Solutions with Gratifying Results*. Chicago, IL: Health Forum, 2008.

Leebov, Wendy, and Gail Scott. *Service Quality Improvement: The Customer Satisfaction Strategy for Health Care*. Lincoln, NE: Authors Choice Press, 2007.

Lill, Marianne M., and Tim J. Wilkinson. "Judging a book by its cover: descriptive survey of patients' preferences for doctors' appearance and mode of address." *British Medical Journal* 331 (2005): 1524–1527.

Matsumoto, David, Mark G. Frank, and Hyi Sung Hwang. *Nonverbal Communication: Science and Applications*. Los Angeles: Sage Publications, Inc., 2013.

Montague, Enid, Ping-yu Chen, Jie Xu, Betty Chewning, and Bruce Barrett. "Nonverbal interpersonal interactions in clinical encounters and patient perceptions of empathy." *Journal of Participatory Medicine* 5 (August 14, 2013).

Morse, Jennifer S. "Improving Patients' Satisfaction through Positive Communication." *Cataract and Refractive Surgery Today* (April 2009): 102.

Morgan, John and Martin *Brenig*-Jones. *Lean Six Sigma For Dummies.* Chichester, West Sussex, England: John Wiley & Sons, Ltd., 2009.

Newberg, Andrew, and Mark Robert Waldman. *Words Can Change Your Brain: 12 Conversation Strategies to Build Trust, Resolve Conflict, and Increase Intimacy.* New York, NY: Penguin Group, 2012.

Oveis, Christopher, June Gruber, Dacher Keltner, Juliet. L. Stamper, and W. Thomas Boyce. "Smile Intensity and Warm Touch as Thin Slices of Child and Family Affective Style." *Emotion* 9, no. 4 (August 2009): 544–548.

Papp, Eric. *Leadership by Choice: Increasing Influence and Effectiveness through Self-Management.* Hoboken, New Jersey: John Wiley & Sons, Inc., 2012.

Patterson, Kerry, Joseph Grenny, David Maxfield, Ron McMillan, and Al Switzler. *Influencer: The Power to Change Anything.* New York, NY: McGraw-Hill, 2008.

Patterson, Michael M. "Touch: Vital to Patient-Physician Relationships." *The Journal of the American Osteopathic Association* 112, no. 8 (August 2012): 485.

Peters, Thomas J., and Robert H. Waterman, Jr. *In Search of Excellence: Lessons from America's Best-Run Companies.* New York, NY: Warner Books, 1982.

Pink, Daniel H. *A Whole New Mind: Moving from the Information Age to the Conceptual Age*. New York, NY: Penguin Group, 2005.

Press Ganey. *2011 Pulse Report: Perspectives on American Health Care*. South Bend, IN: 2011. Accessed online September 29, 2013. http://www.pressganey.com/researchResources/hospitals/pulseReports.aspx.

Rashotte, Lisa Slattery. "What Does That Smile Mean? The Meaning of Nonverbal Behaviors in Social Interaction." *Social Psychology Quarterly* 65, no. 1 (March 2002): 92–102.

Reason, James. *Managing the Risks of Organizational Accidents*. Burlington, VT: Ashgate Publishing Company, 1997.

Redelmeier, Donald A., Jean-Pierre Molin, and Robert J. Tibshirani. "A Randomized Trial of Compassionate Care for the Homeless in an Emergency Department." *The Lancet* 345 (1995): 1131–34.

Rothenberg, Ken J., Nancy Eisenberg, Christine Cumming, Ashley Smith, Mike Sing, and Elizabeth Terlicher. "The contribution of adults' nonverbal cues and children's shyness to the development of rapport between adults and preschool children." *International Journal of Behavioral Development* 27 no. 1 (2003): 21–30.

Schein, Edgar H. *Organizational Culture and Leadership*. 3rd ed. San Francisco: Jossey-Bass, 2004.

Seuss, Dr. [Theodor Seuss Geisel]. *The Lorax*. New York: Random House, Inc., 1971.

Shaw, Kenneth A. *The Intentional Leader*. Syracuse, NY: Syracuse University Press, 2005.

Sorry Works! Accessed online September 24, 2013. http://www.sorryworks.net/.

Studer, Quint. *Hardwiring Excellence*. Gulf Breeze, FL: Fire Starter Publishing, 2003.

Studer, Quint. *Results that Last: Hardwiring Behaviors that Will Take Your Company to the Top*. Hoboken, New Jersey: John Wiley & Sons, Inc., 2008.

The Beryl Institute. "The Patient Experience." Accessed November 15, 2013. http://www.theberylinstitute.org/?page=DefiningPatientExp

Thompson, Jeff. "Is Nonverbal Communication a Numbers Game?" accessed August 20, 2013. http://www.psychologytoday.com/blog/beyond-words/201109/is-nonverbal-communication-numbers-game.

Zolnierek, Kelly B. Haskard, and M. Robin DiMatteo. "Physician Communication and Patient Adherence to Treatment: A Meta-analysis." *Medical Care* 47, no. 8 (August 2009): 826–834.

www.ingramcontent.com/pod-product-compliance
Lightning Source LLC
Chambersburg PA
CBHW021921170526
45157CB00005B/2124